"GOD BL...

THIS YEAR OF CULTURE"

Creating a Hymn

Reciting a Hymn

Singing a Hymn

Godfrey Holmes

© Godfrey Holmes July 2019
ISBN : 978-0-9934644-0-9

All rights reserved. No part of this publication may be reproduced or transmitted in any form or by any means: electronic or mechanical, including photocopying, recording, or by any information storage and retrieval system- nor may any poem be publicly performed – without the prior permission, in writing, from the publisher. A catalogue record of this collection is available from the British Library

NETHERMOOR BOOKS
"St. Elphin,"
12 North Promenade
Withernsea
HU19 2DP

Telephone : 01964-615258

Contact the Author : godfrey.holmes@btinternet.com

All proceeds of Sale of this Book will go to St. Matthew's Church, Owthorne - **or** *the alternative Church or Chapel selling complimentary Copies. However, any opinions the Author expresses in the Hymn itself - or in the accompanying Text - are his and his alone.*

Dedicated

to

Revd. Martin T. Faulkner

<< The Pulpit is very exalted;
The Job very humble;
The Benefice very wide;
The Audience very narrow;
The To-Do List very long;
Time very short -
Yet the Incumbent: very resourceful >>

CONTENTS

The Need for a Hymn of Celebration : p. 1

Getting Started : p. 2

Developing my Hymn of Celebration: p. 6

Composing Verse Number One : p. 7

Composing Verse Number Two : p. 10

Composing Verse Number Three : p. 11

Composing Verse Number Four : p. 13

Composing Verse Number Five : p. 14

Composing Verse Number Six : p. 16

Composing Verse Number Seven : p. 18

Composing Verse Number Eight : p. 20

CONTENTS... Continued

Making a Decision on Performance : p. 22

Deciding on the Other 75 Places : p. 25

The Public's Reaction to Outside Performances of the Hymn of Celebration : p. 26

Pop-Up Performances of the Hymn of Celebration Nos. 1-77 : p. 29

Pop-Up Performances of the Hymn of Celebration Nos. 78-144 : p. 39

Pop-Up Performances of the Hymn of Celebration Nos. 145-222 : p. 49

A Surprising Finish to my Outdoor Recitations : p. 60

Conclusion : p. 62

Text of the full Hymn itself : p. 65

Origins of Hull's Year of Culture

UK City of Culture happens once every four years, concentrating on one location in the UK : to promote arts and culture; twelve months of celebration and regeneration. The initiative is administered by the Department for Culture, Media and Sport, all building on the success of Liverpool's year as European Capital of Culture 2008, which had significant social and economic benefits for the area. The inaugural holder of the award was Derry~Londonderry in 2013–2016.

After Derry~Londonderry in 2013 the next UK City of Culture was scheduled for 2017. Officials from Aberdeen stated they would bid for the title - as did advocates of Dundee, Derby, Leicester, Plymouth,[Stoke-on-Trent, & Swansea. Portsmouth and Southampton made a joint bid for the 2017 title.

In June 2013 a shortlist of four bids: Dundee, Hull, Leicester and Swansea Bay was announced. The winner of the 2017 title was announced on 20 November 2013 and *Hull* was chosen. TV producer Phil Redmond, who chaired the City of Culture panel, said Hull was the unanimous choice because it put forward "the most compelling case based on its theme as a City coming out of the shadows" On 31 July 2014, Martin Green was announced as chief executive of the team. Green was previously head of ceremonies for the 2012 Summer Olympics, and organised the 2014 Tour de France Grand Départ ceremony in Yorkshire.

****

THE NEED FOR A HYMN

It was not self-evident that Hull's Year of Culture actually *needed* a *Hymn*.

After all, the City Council, its Departments, its Schools, its leisure & shopping opportunities were moving - drifting? - towards Secularism.

Might Religion "spoil" the Year of Culture - like a troublesome Guest arriving late at a Party to which she has not been invited?

On the other hand, other big Occasions *have* merited a special Hymn: Victory in Europe (1945) the Coronation (1953), the Queen's Silver Jubilee (1977), the Millennium (January-December 2000), Telford New Town (2012) - or the 50th. Anniversary of the Aberfan Disaster (2016). Just one or two of these Compositions sank without trace.

So Hull, 2017, was certainly an attraction for any local Poet's celebratory - & rhyming - pen: especially as the City has been home to so many famous Poets & Authors, from Andrew Marvell to Philip Larkin; from Alan Plater to Val Wood. Actors & Musicians, too, were inspired by Hull's renewed fame - plus the Year of Culture's concomitant newsworthiness.

GETTING STARTED

Especially if the Composer is *not* a Musician, it is vital the hymn-writer has constantly in mind :

1) <u>A borrowed tune or rhythm</u> - for guidance *from beginning to end* of Composition. I tend to borrow old-fashioned Methodist hymn tunes from my youth - or folk songs or ballads I have heard since. I would find "cold" Composition much more difficult.

For my *Hymn* for Hull's Year of Culture : the thumping-highly memorable - tune used for
    <<*The Church's One Foundation*>> (*Aurelia*)
was an obvious choice : very familiar, very syllabic - no awkward twists or refrains.

7-6-7-6-7-6-7-6 is not the *commonest* metre for a hymn - but has the great advantage of *length* : eight lines furnishing far more room for developing a subject than either 4 or 6 lines.

Even then, each Composer is responsible for stresses: stress either recurring at irregular intervals - or strictly alternate - like the movement of oars in a rowing boat: *Compare*: "The CHURCH's one found-A-tion IS Jesus Christ our Lord..." *and* "Yet SHE on EARTH hath U-nion with GOD the THREE in ONE..."

or: " DING-Dong-MER-ril-Y-on-HIGH; In-HEAV'N- the-BELLS-are- RING-ing..."

That understood, imagined - or timeworn - stress is not *automatic*. The Singer is left free to introduce it, adhere to it, *or withhold it*, for reasons choral, theological, theatrical - or satirical. For instance:
" Mine EYES have seen the GLO-ry of the COM-ing of the Lord! ..." could sound like: "Mine eyes have SEEN the glory of the coming of THE LORD! ..." or " MINE eyes have seen the GLO-RY of the COM-ING of the Lord! ..."

Books & Guides promoting the understanding of Poetry are full of complicated instructions as to how and when stress should be built into verse. I prefer the more pragmatic outcome: *stress is difficult to describe but you'll know it when you meet it.*

****

2) <u>A first line</u>. Hard to believe: a "good" first line opens the way for all the Composition that follows it - even when that line is *not* repeated to commence all succeeding verses.

<< *God Bless This Year of Culture*>>

suited me perfectly. After all, this was to be *a hymn*, not just a poem - though all hymns are, by definition, poetic. Additionally, *God* had to be central not only to the *Hymn* itself but also to the artistry & creativity underlying - and on display throughout - any "Year of Culture." So *God* needed to be the very first word: & this time *not* partially disguised as *Thou, He, One, Rock* or *Almighty*.

****

3) <u>A structure</u>. Arguably, a hymnist could get to the end of his/her hymn *without* a structure. But that Composition might then lose its way.... & meander.

One hymn writer chose the structure of: God at Dawn, God at Noon, God at Homecoming, finally God at Retiring to Bed. Another writer took the breadth, length, depth, height of God's mercy: that was four verses settled outright -plus one verse of Introduction, one of Conclusion.

From the very start - sitting on a village bench prior to a village church Concert - I knew *my* Hymn for Hull, Year 2017, had to devote a separate verse to each of the great expressions of Culture. The question then arises: which are the most important or influential branches of artistry? Which disciplines are "cultural" as opposed to defining hobby or paid employment?

And is *Sport* - or sporting prowess - sufficiently grand, *sufficiently graceful*, to qualify as artistry? Including *Music* arouses no dispute. *Painting*: is also a definite. *Sculpture* qualifies for a verse of its own - because I love sculpture: particularly sculpture created for the great outdoors. *Poetry* proves another definite. And I could not possibly ignore *Drama* : using Stage for the dramatic acting out of a story or episode.

Enough! Enough? Except I could - *perhaps should* - have dedicated one complete verse to *Dance*. Also: what about *Novels*? Where should *Magic* fit in? And surely a verse could concentrate on *Gymnastics* : gymnastics already an ingredient of many a Hull Freedom Festival? But length! Length? The longer my *Hymn*, not necessarily the more accomplished its Composition. Because most Compositions have a *natural* length: a balance exceeded with caution.

That, therefore, is how I achieved an *eight-verse* length - comprising eight relatively long verses - in a context where Church congregations do *not* welcome hymns longer than 5 stanzas of 8-lines, or 7 stanzas of 4 lines: 200 seconds or so of actual singing.

DEVELOPING MY HYMN OF CELEBRATION

<< *A good hymn is the most difficult thing in the world to write. In a good hymn you have to be commonplace and poetical. The moment you cease to be commonplace and you put in any expression at all out of the common, it ceases to be a hymn.* >>
[Alfred Lord Tennyson]

Within 2 hours on that none-too-warm public bench, I had no trouble scripting most of Verse 1, a bit of Verse 2, part of Verse 3, two lines of Verse 4 and several lines of Verse 6.

More important, all the seeds were sown for how my *Hymn* would actually develop... especially the first six first lines. It would appear essential that anybody attempting to compose a hymn of this importance does not try to achieve perfection, verse by verse, straightway. *Perfection* comes in succeeding days - occasionally in succeeding *weeks* - maybe *never*.

Certain verses have to be abandoned as unworkable; then entirely re-written. Counter-intuitively: more satisfaction is gained when Composition and Composer alike face setback than with everything falls too easily into place.

COMPOSING VERSE NUMBER ONE

GOD BLESS THIS YEAR OF CULTURE :
Of skills, invention, art....
Direct our wills more surely
To play the fullest part.
You gave us joy and anguish:
Hopes dashed, ambitions raised....
Lest we in sloth should languish :
Creation's gifts *be praised.*

This verse has to be both Introduction and fanfare; also a foundation for all succeeding verses. Moreover, it is essential I do not, unintentionally, *copy* hymns remembered from my childhood.

I inject a positive note by including my *audience* as participants in & contributors to Hull's *Year of Culture*- not merely observers; second, by placing joy *before* anguish; third by doing *the opposite* in line 6: placing hopes dashed earlier than ambitions raised.

My thanksgiving to God : *"You gave us joy & anguish"* is actually a memory of my singing: *"He sends the snow in Winter, the warmth to swell the grain....All good gifts around us are sent from Heaven above...."*; & *"We owe Thee thankfulness & praise, Who givest all,"* in childhood Harvest Festivals.

When & where I also sang :

> *<< He gave us eyes to see them,*
> *And lips that we might tell*
> *How great is God Almighty,*
> *Who has made all things well! >>*

Is *"sloth"* too *obscure* a description of inertia: a relic from Bunyan's *Pilgrim's Progress?* One always has to be cautious when singing or praying a *peculiar* word or peculiar concept- lest its sheer singularity jars and causes the singer/worshipper to question its inclusion. Unfairly, that means hundreds of perfectly good words are disqualified on the grounds of un-sing-ability. Remember how *this* example of startling modernity was greeted:

> *<< God of Concrete, God of Steel;*
> *Lord of Piston, Lord of Wheel >>*

By way of illustration, here is an alphabetical snapshot of words almost unemployable in the lexicon of worship - however innocuous those same words in a novel or play or sermon:

Authority, Blame, Concession, Deceit, Ethereal, Frantic, Generosity, Helpful, Intrepid, Justification, Kick, Levity, Moronic, Neutral, Optimum, Preventive, Querulous,

Remainder, Suspicion, Terrific, Upstage, Withheld, Youthful, Zionism.

How fortunate I am that *"languish"* rhymes with *"anguish."* Neither of these words is especially common in everyday speech - but that rarity is amply offset by the sheer pleasure of singing that is *swish*.

Significantly in *Verse One*: I am able to release line 3 from having to rhyme with line 1: a feature I usefully retain for each succeeding verse. I also manage to reserve lines 1 to 4 for people's affiliation ; lines 5 to 8 for their *emotion*.

A few leading versifiers effortlessly "end" a line or verse mid-clause, mid-sentence. I choose the more conventional format of not permitting half a thought or half a description to hang in mid-air. And in the case of *"direct our wills to play the fullest part,"* I insert *"more surely"* as parenthesis.

****

COMPOSING VERSE NUMBER TWO

GOD BLESS THIS YEAR OF MUSIC:
Of solo, chorus, band....
Alert our ears, distinctly,
Each score to understand.
You gave us fine musicians ;
The drum, horn, cello, flute :
Notes chosen with precision;
Performances astute.

Music is so important an ingredient of Culture, Music not only deserves a verse of its own, it also occupies key *Verse Two*. Nor does it matter what type of Music might or might not emerge during the *Year*.

In reality, folk & jazz in clubs and in pubs emerge; outdoor singer-guitarists; bands; pop & rock both indoors & on an outdoor stage; classical concerts in City Hall, also in churches; quartets, quintets, choral societies on street corners & in Victoria Square. So *Verse Two* has to be *imprecise* concerning genre - while still urging each future, paying or non-paying audience member to think about Performance booked or Performance impromptu.

My Petition : " *Alert our ears distinctly...*" is actually an echo of the *"Direct our wills more surely..."* of *Verse One* - just as *"You gave us fine musicians"* echoes the

"*You gave us joy & anguish*" of *Verse One*. Of interest, I am not including *every* instrument that is played in concert. Therefore: "*The Drum, Horn, Cello, Flute...*" are instruments that not only *scan* well but prove indispensable whenever talented musicians make their entry to stage.

In *Verse Two* I find "*precision*" rhymes not with an exactly pronounced "decision," but with "*musician*": not really to its detriment - because most poets resort to ingenious half- or three-quarter- rhymes. "*Flute*" appears a perfect match for "*astute*" - but a purist would say you cannot couple *shoe* [-oo] with *blue*.

****

COMPOSING VERSE NUMBER THREE

GOD BLESS THIS YEAR OF PAINTING :
Of landscape, portrait, view....
You gave us oils and pastels ;
Ink, chalk - and subtlest hue.
You nurtured clear perspective :
The illustrative urge ;
Each camera shot elective....
Our visionary surge.

"*Painting*" is an odd way to describe the laborious process of creating a work of art indisputably good enough to hang in Hull *Ferens' Art Gallery*. After all, we paint doors, window frames, walls, benches. And that is also the problem with the word : "*Art*." If a child opts to study Art at 'A' Level, he or she expects to be presented with an array of easels, tubes, brushes, pencils, canvases - with a camera too - but not sandpaper, a chisel, a scraper, a violin, railway timetable - or bow and arrow. For Art is too broad to define or describe.

But "Painting" it has to be. "Landscape" & "Portrait" are of course the two ways of arranging a sheet of A4. "You gave us..." is an intentional repeat of those exact words in *Verse Two*. [Note my contemporary abandonment of the address : "Thou"]. "Ink" & "chalk" I include as useful alternatives to oils.

In lines 5-8, I turn from tools to motivation, from colourist to bystander: creativity, but strictly *optional* creativity when contrasted with installing gas or electricity, or selling food. Retention of the words "us" and "our" continues to aid inclusivity. And I am fortunate "urge" & "surge" rhyme so easily.

****

COMPOSING VERSE NUMBER FOUR

GOD BLESS THIS YEAR OF DRAMA :
Of script and costume....*Dance*....
Our love of masquerading;
Deft change of voice and stance.
Each actor's declamation -
Your mission-fields the stage -
Draws loudest acclamation....
Up risen from the page.

I am very fortunate in *Verse Four* to have no quarrel with the word *"Drama"* nor with drawing in the Dance that might on another day have been found in a verse all of its own. I am also blessed with no less than 3 four-syllable words : *"masquerading," "declamation,"* & its cousin: *"acclamation."*

All versifiers have difficulty with 4-, 5-, & 6-syllable words lest they run out of available syllables; also lest they throw scansion overboard. For instance: *"cosmopolitan"* is a very handy peephole into both city-living *and* the Russian Orthodox faith. But no way does it sit so easily as Charles Wesley's brilliant 6-syllable: "inextinguishable" to describe the blaze of the Holy Spirit in *"O Thou Who Camest from Above..."*

Continuing with the theme of *fortune,* all the imagery for lines 5-8 of *Verse Four* comes to me, complete, on a double-decker bus a few days after I embark on

Composition. How I manage to reflect Jesus' Resurrection in words "leaping from the page," I shall never know. All I *do* know: *all of life* is drama - and the Stage really is as universal as Empire.

Just hear the clapping!

****

COMPOSING VERSE NUMBER FIVE

GOD BLESS THIS YEAR OF SCULPTURE
Of object, statue, plate:
Form shaped by flame or chisel,
Symbolic and ornate.
You gave us wood, clay, metal,
The craftsman's steady hold;
Ere final glaze or fettle
Imagination ***bold.***

The only word to describe this Verse is "unachievable." Despite my intense interest in Sculpture from Battersea Park, Year 1966 onward, this Verse refuses to be written - therefore refuses to be re-written. It desperately requires a central plank which turns out to be the *"Form shaped by flame or chisel, symbolic and ornate."*

Because, arguably, Sculpture is one of the most supreme - as well as one of the most technically difficult - aspects of Culture. Undoubtedly, we require Busts & Statues: lots of them. We also need Ceramics both on the table and in the china cabinet. So why is Sculpture so difficult to include in a hymn?

Maybe it is envisioning what the final product will look like: triumphalist or minimalist, temporary or permanent, glittery or matt? All through the Old Testament there is debate as to whether God, in particular, can adequately, respectfully, be represented in idolatry? ... the act of idolatry breaking almost every Commandment!

To muddy the waters further, controversy surrounds many a *finished* Statue. *"The Meeting Place"* in a beautifully re-modelled *St. Pancras Station* is a success; Cristiano Ronaldo in Madeira less so.

Here, lines 5-8 do not come so easily - but at least I can resort to an exact repetition of *"You gave us..."* as in line 5 of *Verses One, Two,* & the eventual *Verse Seven*. Here, I recall three of the most popular materials at the disposal of a sculptor: wood, clay & metal; for good measure thanking God, too, for the "craftsman's steady hand." I have to doubt the inclusivity of the word *"craftsman"* - as opposed to "*craftswoman*" or "*craftsperson*" - but craftsman it has to remain in the hope both familiarity and flow.

Little time elapses before I come to cherish that "final glaze or fettle": itself an acknowledgement of process and perfection. The painter touches up. The musician fine-tunes. The dancer adjusts her choreography. But nowhere is finish, *completion,* as vital - maybe as irreversible - as in Sculpture.

****

COMPOSING VERSE NUMBER SIX

GOD BLESS THIS YEAR OF VERSES :
Of poetry and rhyme....
Each tantalizing stanza -
Delivered in strict time;
Quotation loose or metric -
To memorize, recite;
Lines soothing, or electric :
Rehearsed through silent night.

Verse Six: another fairly "easy" one to compose - partly because Poetry is an immediately familiar, & by definition legitimate, facet or by-product of Culture.

General Poetry is no simpler in the making than hymnody. Most poets are perfectionists: so will not let their verse see the light of day before it presents well and reads engagingly. Word pictures might appear precious, meanings incomprehensible - but the readers of Poetry must stand in awe of the Poet.

It follows that the best verse is tantalizing: it tantalizes not only the reader but also the wordsmith - and ultimately it tantalizes itself. The words want to escape the rigidity of form but cannot. So most words within most Poetry are left dangling. They sleep until awoken by understanding.

I'm glad my Verse about God blessing a Year of Verses is to be *"delivered in strict time"* because: were it or its neighbouring Verses ever bungled or trashed through poor, halting, or uncomprehending delivery, they would not make the impression, or evoke the reaction, I intend for them.

That preoccupation with function, process and entertainment value spills over into lines 5-8, making them a continuation of lines 2,3 & 4. Some Poetry will indeed be memorized: more difficult to do than first imagined. That leaves ample room for *recitation*: reading a text out loud, at the same time looking *down* for guidance; *in front* for appreciation and feedback. In line 7, I underline the fact that some

Poetry is soothing enough to help & to reassure both the giver and receiver.

"*Silent night,*" silent night: echoes of that famous Christmas Carol of the same title! And why is Poetry rehearsed *at night*? By the Composer to get it right; by the insomniac to get back to sleep; by the performer to brush up.

I can do a lot more with this Verse but choose not to. That is how elusive versifying is.

****

COMPOSING VERSE NUMBER SEVEN

GOD BLESS THIS YEAR OF VENTURE :
By dreaming, to conceive ?
Glad days of exultation;
Displays at morn or eve.
You gave us flair and passion
Exciting sense and sight....
Beyond all transient fashion:
Each person's worth, de*light*.

This Verse marks the end of specifics, a return to generalities. "*Venture*" is a lovely word: so much more descriptive of endeavour than simple *ad*venture. When we venture forth we are uncertain as to outcome - yet accept, even cherish, the risk factor that homecoming might be heavy with disappointment.

Or a venture could be a new assignment, a new responsibility, a new opportunity : exciting indeed. And Hull's *Year of Culture* was to be nothing if not exploratory, ingenious.

Whoever bid, whoever angled, for Hull - as opposed to a dozen other contenders - to get the honour was dreaming a dream, maybe *fulfilling* a dream. Luckily, the Year *was* to be a dream come true. Because there was a sufficient pool of creativity, plus a feeder stream of ideas, to give the long-anticipated Year some oomph.

"Glad" & *"gladness"* both are words used relatively rarely in hymnody. *"Let us with a gladsome mind praise the Lord for He is kind,"* & the anthem: *"I was GLAD when they said unto me..."* come immediately to mind. Here I use *"glad"* to describe the 365 days of 2017 - when there would indeed be "displays at morn & eve" : *eve* very handily rhyming with *conceive*.

Then I return to those familiar words from *Verses 1,2,3,&5*: *"You gave us...."*: this time *"flair & passion"* - because without passion there can be no great music, paintings, photography, drama, dance, ceramics or sculpture. Sealing this immense contribution is God's gift of *sense* & *sight*: a none-too-disguised half quotation of the words: *"Linking sense to sound & sight..."* from the hymn so amazingly rearranged by John Rutter: *For the Beauty of the Earth...*

At the end I am asserting - or alluding to - the transience of many works of art. Most of Rossini's operas have been rarely performed; many pop songs don't stay top of the charts; plays get forgotten or removed from stage earlier than planned; outstanding dance sequences might not be repeated later in the series; china gets broken; statues rust. Maybe I am predicting the fate of certain less pleasurable, less skilled, components of Hull's *Year of Culture*.

I finish *Verse Seven* on a more positive note: recognizing each person's "worth"; also the "delight" they might both give and receive through throwing themselves into whatever is happening - even standing stark naked beneath the iconic Humber Bridge - in a rejuvenated, reinvigorated Hull.

****

COMPOSING VERSE NUMBER EIGHT

TO HULL, GOD OF ALL CULTURE :
We varied talents bring:
This year - and in the future -
To serve our glorious King;
So we and all your people -
Blest common purpose found -
May make this chosen City
Your Kingdom's hallowed ground.

If *Verse* 5 [Sculpture] is almost impossible to construct, Verse 8 of the Hymn to celebrate Hull in 2017 is 10 times more challenging!

How can I sum everything up? How can I bring the different ideas behind Culture together in one place? My immediate hope is to make reference to the extensive bombing of the City in the Second World War: a catastrophe hidden from the British public at the time and for many years later. Rebuild! And how better to rebuild than through Art & Architecture?

Hull got the Architecture aplenty, Post-War - but it was of only mediocre quality. Many new cinemas, shops & skyscrapers were either ordinary or ugly or both. And bombing is rather a *down* way to end an *up Hymn*.

Eventually, Pam, a fellow student from Withernsea High School past comes to my rescue and writes Verse Number 8 much as it materializes. The Verse is *still* controversial: first because an untutored listener might think God is called "*Hull*" much as in some parts of the world God is called anything from *Abba* to *Adonai*, from *Elohim* to *Haneunim*. One Hebrew name for God : YHWH is so holy it cannot be said out loud!

Then the words "*King*" & "*Kingdom*" are both irredeemably patriarchal. I personally have biggest difficulty with the concept of God as "*Father*." Not only is fatherhood an analogy: the Church as family; calling God *Father* is very forbidding, very haunting, for adults who have very negative childhood memories of relating to one particular father, one particular stepfather, one particular grandfather.

Pam excels in sounding upbeat: & not just for the *Year* ending December 31st., 2017. She looks to *the future*: a future of service; a future of commonality; a future of joint enterprise; best, a future of artistic holiness - art sacred & the sacred, art.

Verse 8, therefore the *Hymn* and the *Year*, ends on "*hallowed ground*" - very Moses!

****

MAKING A DECISION ON PERFORMANCE

I knew the first public performance of my *Hymn* would, appropriately, be at my nearest Parish Church: *St. Matthew's* Owthorne. I also knew I needed to give a 5-minute Introduction to this rendering: how I came to be in or near Hull; how I came to study & embrace

Hymnology; also, & most relevant, what gave me the idea to write: *"God Bless This Year of Culture..."*

Concurrently, the *Hull Daily Mail* ran the story behind the *Hymn*. Then a Lottery: the unknown. Will other Churches & Chapels take up the *Hymn*? Maybe. More certainly, *Hull's Year of Culture* is secular: some would say *militantly* secular. The entire Year is to be *God-free* lest any passers-by, any ethnic minorities, any performers, any fragile school-teachers, any "stakeholders" get agitated or upset.

Therefore: literally at the last minute - & two hours before I am due to talk about my *Hymn* - I decide to take my *Hymn of Celebration* out on the road, reciting it wherever I could.

I pay no heed to *how* this would be done without a car - nor *where* exactly; nor what the public's response might be to outdoor recitation. All I know: it will give *"God Bless this Year of Culture"* a definite life; a pervasive presence through the Year.

77 Pop-ups is the number I talk about. Over 12 months, that sounds do-able. 77 is the approximate number of outstanding buildings or statues in Hull [or so I imagined]. Additionally 77 is mentioned in Scripture: not merely as a multiple of 7 but also as the number of times we might forgive someone who has wronged us: alternative reading, 490. Jesus was good on exaggeration to make a point!

Having pledged that first Congregation 77 outings: before January 2017 was further spent, I plan two "book-ends," 2 outer limits: my *first* recitation to be on Humber Bridge Western walkway - facing land - on Tuesday January 31st; my *last:* the 77th ? on Humber Bridge's *Eastern* walkway - facing the Sea - on Friday December 1st.,2017.

A 10-month span: missing out Christmas & Advent - when Hull's *Year of Culture* will be virtually spent in any case. How about those *other* 75 appearances?

Straightway, I decide to set these parameters:

*** No prior notice - unless essential;
*** No pouring rain or driving snow;
*** No fancy dress;
*** No delegation to anybody else;
*** No changing my mind half way through;
*** No trespassing;
*** No obstruction;
*** No fundraising - nor any acceptance of donation;
*** No conversation during recitation;
*** No movement of body or feet during recitation;
***No actual *singing* of the Hymn outdoors;
***No learning off-by-heart.

This list sounds forbidding - and *very* negative. But public performance is a privilege: so it demeans the exercise - and God - to act the fool; or to be blown away by the moment. Learning off-by-heart, for instance, sounds very sensible - but comes unstuck straightway one line is lost, then the next, then the next. "Obstruction" is a particular obstacle: literally. A performer can be moved on for obstruction, at worst pushed to the ground or arrested.

After all the things *not* to do came the positives: recording the exercise for posterity. That entails at least two photographs of each recitation: one of the newly-printed *Hymn* fluttering in front of the chosen or accidental venue; the second of me actually at that venue, preferably holding the *Hymn* [a "selfie"].

Selfies are of course an act of vanity. So how to reconcile: "Glory to Me" against "Glory to Thee"? But *absence* of those selfies [occasionally a properly-framed photo by a member of the public] will risk the charge of falsehood. Modern digital cameras routinely revert to a practice which had fleeting popularity from 1905-1930 following the availability of the Kodak-Eastman *Brownie*: *dating each shot.*

****

DECIDING ON THE OTHER 75 PLACES

Certain places such as the *City Hall*, *Ferens*, *Paragon*, *Hull Kingston Rovers*, *Street Life*, the *Deep*, *William Wilberforce, Holy Trinity* Parish Church- as it was then- the *New Theatre*, the *Truck* & *Guildhall* are all givens: boosted by the entrances to lots of covered shopping malls: definite, but that is a mile short of 75!

Because Hull's *Year of Culture* is supposed to draw in *the region*, as far as the North Lincolnshire that used to be called South Humberside; as far as the coastal alluvial deposit called Holderness; also as far as Howden & Goole to the West, the Yorkshire Wolds to the North, I can afford to spread my wings from Spurn Head to Bridlington Harbour without throwing my net so wide as to catch the whole of the North!

After due consideration of *importance* in the landscape or townscape, my rule-of-thumb becomes: does this place in any way *"face"* the City of Hull or draw strength from it? 75 now looked rather more achievable.

****

THE PUBLIC'S REACTION TO OUTDOOR RECITATIONS OF << *A HYMN OF CELEBRATION*>>

On the afternoon of my first expedition to the Humber Bridge, I sit rather nervously in the Humber Bridge Board's Garden of Remembrance - which I have no prior clue even existed - wondering what exactly I am committing myself to.

I need not have worried. For I am leaving out of all my calculations one technological device I have never used: the ability to talk over a mobile phone without holding the handset alternately close to mouth or ear.

Somebody, somewhere, must have encountered this phenomenon more than I have - because at a sweep this remote control, this remote microphone, abducts, overtakes, the surprise element in the majority of my recitations.

Why turn a hair, why cross the road, when it's only a bog-standard smart-phone? Add to this predictability, maybe *surprise* on its own is less common - therefore less commented upon?

I do genuinely *fear* the exposure of some venues. The early morning ones are easy. I cannot really be sent on my travels by a security guard performing outside *Hull Magistrates' Court* at 7am - nor face an angry publican outside an ancient inn at 9am. That leaves a really scary appointment: *St. Stephen's* forecourt at 6pm on a Saturday.

That is the only occasion youths mock me, so that they ask for a professionally-printed Copy, then screw or tear it up straightway. Bemused workmen at other venues are really on-side as are most passers-by in most places. Some ask questions: especially a party visiting *East Riding Museum* - whilst others offer to take the photograph or take the *Hymn* back to their church or chapel and sing it to one of many alternative tunes I was always willing to signpost: "*O Jesus I have Promised...*" [4 versions], "*Thy hand O God hast Guided*" [Thornbury] - or very melodically: "*All Glory, Laud, & Honour...*"

I am gratified to survive successive visits to *Queen Victoria Square* - when the *Blade* looks everso busy. But I do have difficulty with *Poppies* : the stream of poppies flowing from *Hull Maritime Museum*. Here the Artist insists on "No Photography!"... and there are Guards to enforce this preference. I get round this edict by doing my recitation on nearby *King Edward Street*, with *Poppies* in the background.

My saddest - most disappointing - moment is, ironically, at *Hull University*, very early on in my series of Recitations. Unexpected this. Because I have taken inordinate care to obtain prior permission from the Registrar's Office who are fully acquainted with Date, Time, Intent, Duration. I even clock in. But one or more Moslem(?) students apparently take exception to any mention of a Christian God... thus wheel in Security.

Security nobble me while eating my sandwich on a bench near the *Brynmor Jones Library*, outside of which I have performed about half-an-hour earlier. And Security like neither me - nor my explanation. They do not even recognize their own Registrar! My job longsince done, I leave Campus without further protest. At least I know, first-hand, what being "no-platformed" means.

In defence of other Security firms, no other Guard or Guards give me grief - *anywhere* - nor stand in my way, nor move me on .... *at any time.*

77 POP-UP PERFORMANCES OF
<< *THE HYMN OF CELEBRATION* >>

1) HUMBER BRIDGE.... WEST WALKWAY

2) HULL ROYAL INFIRMARY.....
   THE JOHN ALDERSON STATUE

3) SATURDAY MARKET, BEVERLEY

4) HEDON MARKET PLACE

5) PRINCES QUAY, HULL .... THE GANGWAY

6) QUEEN'S GARDENS, HULL ... THE FOUNTAIN

7) OUTSIDE BEVERLEY MINSTER

8) OUTSIDE GUILDHALL, HULL

9) OUTSIDE HULL MAGISTRATES' COURT

10) *ST. MARY'S* CHURCH LOWGATE

11) SUFFOLK PALACE [GPO / WETHERSPOONS] HULL

12) OUTSIDE HULL HOLY TRINITY CHURCH

13) THE OLD [BOYS'] GRAMMAR SCHOOL,TRINITY SQ.

14) GRAND ENTRANCE TO TRINITY HOUSE, HULL

15) MARKS & SPENCER, WHITEFRIARGATE, HULL

16) OUTSIDE FERENS' ART GALLERY, HULL

17) QUEEN VICTORIA STATUE, HULL

18) [OUTSIDE] CITY HALL, HULL

19) THE CENOTAPH, FERENSWAY, HULL

20) THE PHILIP LARKIN STATUE
    CONCOURSE OF HULL PARAGON
    RAILWAY STATION

21) ALBEMARLE MUSIC CENTRE, FERENSWAY, HULL

22) HULL TRUCK THEATRE, FERENSWAY, HULL

23) BEVERLEY RAILWAY STATION

24) FLEMINGATE SHOPPING MALL, BEVERLEY

25) GUILDHALL, BEVERLEY

26) THE TREASURY HOUSE, BEVERLEY

27) MARITIME MUSEUM, HULL

28) THE BLADE, QUEEN VICTORIA SQUARE, HULL

29) BANKS OF THE RIVER HULL, DRYPOOL

30) OLD HOUSE PUBLIC HOUSE, WHITEFRIARGATE

31) MECCA BINGO / CECIL CINEMA. FERENSWAY

32) ELWELL'S PICTURE OF BRICK BRIDGE, SWINEMOOR: DISPLAYED REGISTER SQ., BEVERLEY

33) *TOLL GAVEL* METHODIST CHURCH, BEVERLEY

34) [OUTSIDE] *ST. MARY'S* PARISH CHURCH, BEVERLEY

35) NORTH BAR WITHIN, BEVERLEY

36) BARTON-UPON-HUMBER RAILWAY STATION

37) GRIMSBY : FISHERMEN'S MEMORIAL SCULPTURE

38) [OUTSIDE] *GRIMSBY MINSTER*

39) THE PUNCH BOWL INN, PRINCES QUAY, HULL

40) PARAGON RAILWAY STATION :
    ANLABY ROAD GRAND [SIDE] ENTRANCE

41) THE TOWER CINEMA, ANLABY ROAD, HULL

42) THE PERFORMING ARTS ACADEMY
    ANLABY ROAD, HULL

43) *BRYNMOR JONES* UNIVERSITY LIBRARY
    HULL UNIVERSITY CAMPUS, COTTINGHAM ROAD

44) HULL UNIVERSITY STUDENTS' UNION BUILDING
    HULL UNIVERSITY CAMPUS, COTTINGHAM ROAD

45) UNIVERSITY MAIN ENTRANCE, WITHIN SIGHT OF VENN BUILDING, COTTINGHAM ROAD, HULL

46) *ST. JOHN* COMMUNITY CHURCH, NEWLAND

47) CLOUGH ROAD SWING BRIDGES, HULL

48) GRIMSBY RAILWAY STATION

49) HORNER'S CINEMA & PUB, ANLABY ROAD, HULL

50) LILY BILOCCA MURAL, ANLABY ROAD, HULL

51) SIDE ENTRANCE OF PROSPECT CENTRE, HULL

52) HULL CENTRAL LIBRARY, FORMER ENTRANCE

53) AMY JOHNSON STATUE, GEORGE STREET, HULL

54) *HEAVEN & HELL* NIGHT CLUB, HULL

55) DISTINCTIVE KCOM WHITE TELEPHONE BOXES, PARAGON STREET, HULL

56) ROWLAND HOUSE, PRINCES QUAY, HULL

57) POPPIES ART INSTALLATION, QUEEN VICTORIA SQUARE, HULL

58) MISSION CHAPEL & PUBLIC HOUSE, OLD HULL

59) OLD SUGAR MILL, PRINCES QUAY, HULL

60) RE-POSITIONED ANDREW MARVELL STATUE, TRINITY SQUARE, OLD HULL

61) ZEBEDEE'S YARD, OLD HULL

62) CARMELITE PRIORY , OLD HULL

63) SPURN LIGHTSHIP, HULL MARINA

64) CELEBRATION OF THE EUROPEAN UNION HULL MARINA

65) HOLIDAY INN, HULL MARINA

66) HEWN MONOLITH SCULPTURE, HULL MARINA

67) *HMS DIANA* (DUNKIRK RESCUE) HULL MARINA

68) STATUE CELEBRATING MASS NORTH EUROPEAN EMIGRATION THROUGH HULL TO LIVERPOOL & NORTH AMERICA, HULL MARINA

69) FRUIT MARKET, HUMBER STREET, HULL

70) HAULAGE MACHINERY DISPLAY, PRINCES QUAY

71) MALTINGS BREWERY, BOND STREET, HULL

72) PEARSON PARK GRAND ENTRANCE, HULL

73) PEARSON MEMORIAL, PEARSON PARK, HULL

74) *ALL SAINTS'* PARISH CHURCH, GREAT DRIFFIELD

75) GREAT DRIFFIELD METHODIST CHURCH

76) MERCHANT NAVY MEMORIAL, SAVILE ST., HULL

77) VICTORIA PIER, OLD HULL -
LANDING STAGE FOR HUMBER FERRIES TILL 1981

****

EXTENDING THE NUMBER OF OUTDOOR RECITATIONS OF << *THE HYMN OF CELEBRATION* >> Pop-Up Performances 78 to 144

78) ALEXANDRA PARK, HULL :
CHILDREN'S PLAYGROUND

79) THE OLD CITADEL, OLD HULL

80) THE OLD WINDING ENGINE, OLD HULL

81) ALEXANDRA DOCK BASIN

82) THE OLD CUSTOM HOUSE, OLD HULL

83) THE *BLUE BELL* INN, OLD HULL

84) THE CHARLES WILSON STATUE, LOWGATE, HULL

85) HOLY TRINITY SCHOOL CHAPEL, ZEBEDEE'S YARD, OLD HULL

86) HULL MARINA : THE LOCK GATES

87) KING BILLY STATUE, MARKET PLACE, OLD HULL

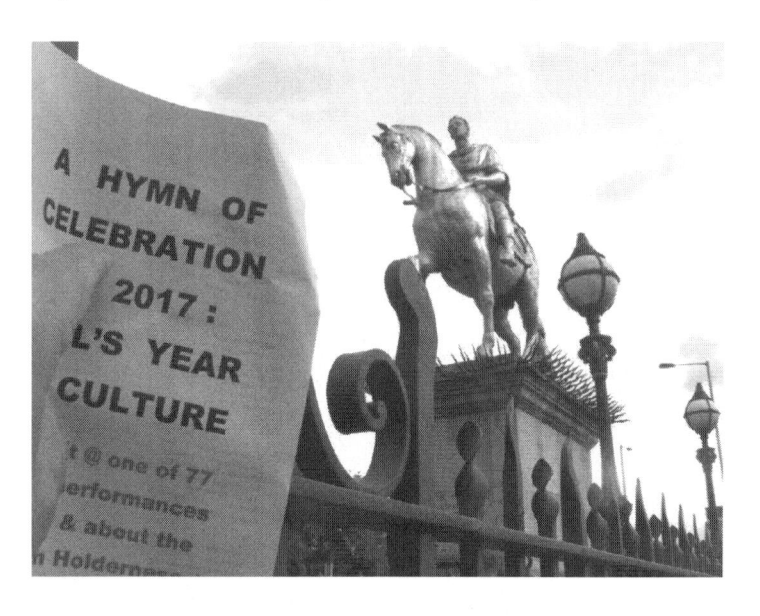

88) MULTI-STOREY CAR PARK, LOWGATE, HULL

89) PREMIER INN, THE DEEP, HULL

90) OLD WOODEN BRIDGE OVER RIVER HULL

91) HULL FLOOD BARRIER

92) OLD HALFPENNY BRIDGE OVER RIVER HULL

93) SPEAKING PULPIT, FRUIT MARKET, HULL

94) LOCKGATES OF HULL MARINA

95) THE OLD FLOUR MILL, DRYPOOL, HULL

96) OLD *WATERLOO INN*, GT.UNION STREET, HULL

97) NORTH BRIDGE HOUSE, BESIDE RIVER HULL

98) NORTH BRIDGE OVER RIVER HULL

99) HULL COLLEGE

100) WILBERFORCE MEMORIAL, HULL COLLEGE / QUEEN'S GARDENS, HULL

101) SIEMMENS NORTH SEA ENERGY FACTORY, HULL

102) EARLE'S SHIPYARD, HIGH STREET, HULL

103) SUNKEN VESSELS NEAR VICRORIA DOCK, HULL

104) PETTINGEL's MAP near VICTORIA DOCK, HULL

105) SCULPTURE in VICTORIA PARK, HULL

106) HARTLEY's BRIDGE, VICTORIA DOCK, HULL

107) THE SUN DIAL beside VICTORIA DOCK, HULL

108) THE EGYPTIAN ENTRY to VICTORIA PARK, HULL

109) WOODEN STATUE within GROUNDS of DEMOLISHED *ST. PETER's* CHURCH, DRYPOOL, HULL

110) PIER TOWERS, WITHERNSEA

111) ACOUSTIC TENT, WIVFEST, WITHERNSEA

112) GIANT PICTURE POSTCARDS, CENTRAL PROMENADE, WITHERNSEA

113) WITHERNSEA's FISHING COMPOUND

114) OLD MUNICIPAL OFFICES, WITHERNSEA

115) *ALL SAINTS'* CHURCH, TUNSTALL [YORKSHIRE]

116) BEACON for TRANS-PEAK TRAIL, HORNSEA

117) THE SAILING CLUB, HORNSEA PROMENADE

118) OLD RAILWAY STATION, HORNSEA

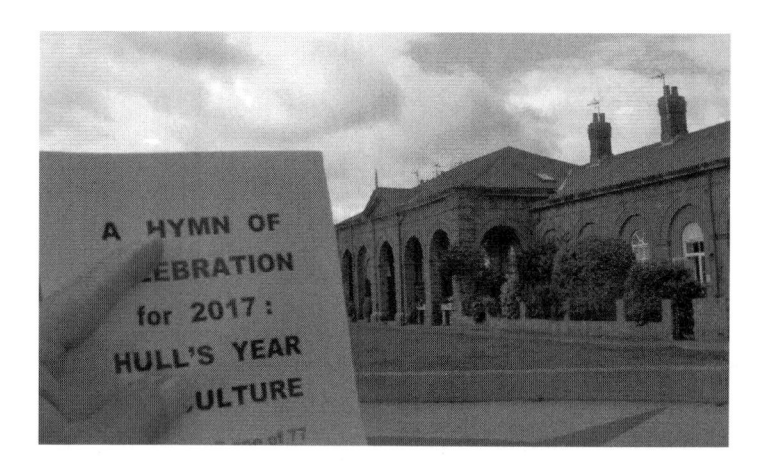

119) LIONS at the entry to MEMORIAL GARDENS, HORNSEA

120) WESLEYAN SCHOOLS / TOWN HALL, HORNSEA

121) *ST. NICHOLAS* PARISH CHURCH, HORNSEA

122) IRON STATUES, on Platform of STEPNEY RAILWAY STATION, BEVERLEY ROAD, HULL

123) *ST. NICHOLAS* PARISH CHURCH, WITHERNSEA

124) METHODIST CHURCH, HULL ROAD, WITHERNSEA

125) LIGHTHOUSE GROUNDS, HULL ROAD, WITHERNSEA

126) *ST. MATTHEW's* PARISH CHURCH, OWTHORNE

127) WITHERNSEA HIGH SCHOOL: myself a Pupil!

128) PRIMITIVE METHODIST / PENTECOSTAL CHURCH, EASTGATE, HORNSEA

129) BETTISON's FOLLY, NEWBEGIN, HORNSEA

130) MILLER's STATUE, BEVERLEY BECK

131) BEVERLEY WHARFE, RIVER HULL, BEVERLEY

132) *ST. NICHOLAS* CHURCH, HOLME CHURCH LANE, BEVERLEY

133) FISHERFOLK STATUE, HULL

134) THE PIER, CLEETHORPES

135) HILTON DOUBLETREE HOTEL, FERENSWAY, HULL

136) HULL DAILY MAIL OFFICES, BEVERLEY ROAD, HULL

137) *ST. JOHN'S* METHODIST CHURCH, OLD BRIDLINGTON

138) *BRIDLINGTON PRIORY* [CHURCH of ENGLAND]

139) THE RAILWAY STATION, BRIDLINGTON

140) *CHRIST CHURCH*, BRIDLINGTON NEW TOWN

141) BRIDLINGTON HARBOUR

142) FORMER QUAKER MEETING HOUSE, OLD HULL

143) *ST. CHARLES BORROMEO* ROMAN CATHOLIC CHURCH, ALBION STREET, HULL

144) DAVID WHITFIELD STATUE, KINGSTON SQUARE, HULL

EXTENDING THE NUMBER OF OUTDOOR RECITATIONS OF << *THE HYMN OF CELEBRATION* >> Pop-Up Performances 145 to 222

145) HULL's NEW THEATRE

146) HULL's HISTORY CENTRE, WORSHIP STREET

147) HULL's FORMER FIRE STATION, WORSHIP STREET

148) KINGSTON COURT, KINGSTON SQUARE, HULL

149) *ST.STEPHEN's* SHOPPING MALL, HULL

150) OLD WHARVES, RIVER HULL, HULL

151) BLAYDES' HOUSE, HIGH STREET, HULL

152) OLD DOCK OFFICES, HIGH STREET, HULL

153) THE DEEP, HULL, from afar

154) DE-LA-POLE STATUE, NELSON STREET, HULL

155) SCANDANAVIAN STATUE, NELSON STREET, HULL

156) PATRINGTON WAR MEMORIAL

157) HULL's MUNICIPAL CEMETERY, HEDON ROAD, HULL

158) HULL PRISON

159) NEWTOWN, SOUTHCOTES LANE, HULL

160) *ST. JOHN'S* CHURCH, SOUTHCOTES, HULL

161) EAST HULL FIRE STATION, SOUTHCOTES, HULL

162) *CHURCH OF THE SACRED HEART*, SOUTHCOTES LANE, HULL

163) LOST VILLAGE OF RAVENSER PUB, SOUTHCOTES LANE, HULL

164) HULL's ORIGINAL SAVINGS' BANK, now The BANK BAR, SOUTHCOTES LANE/ HOLDERNESS ROAD, HULL

165) JAMES STUART STATUE, HOLDERNESS ROAD, HULL

166) HOLDERNESS ROAD METHODIST CHURCH, HULL

167) MOUNT PLEASANT ASDA, MOUNT RETAIL PARK, HOLDERNESS ROAD, HULL

168) CORN MILL HOTEL, MOUNT PLEASANT, HULL

169) [Former] SALVATION ARMY HQ., HOLDERNESS ROAD, HULL

170) RECKITT's LIBRARY, HOLDERNESS ROAD, HULL

171) JORDAN's CAR DEALERSHIP, HOLDERNESS ROAD, HULL

172) THE VAULT NIGHT CLUB, WITHAM, HULL

173) *SAMUEL PLIMSOLL* HOTEL, EAST HULL

174) NAPOLEON's CASINO, HULL

175) *HOLY TRINITY* [BOYS'] ACADEMY, HULL

176) CLUB BIARRITZ, GEORGE STREET, HULL

177) *ST. NICHOLAS* CHURCH, KEYINGHAM

A HYMN OF CELEBRATION

178) PATRINGTON RAILWAY STATION [Closed]

179) PATRINGTON HAVEN CHAPEL & STATUE

180) PATRINGTON HAVEN HOLIDAY CAMP

181) DUNEDIN HOUSE HOTEL, PATRINGTON

182) PATRINGTON METHODIST CHURCH

183) *ST. PATRICK's* CHURCH
<< THE QUEEN OF HOLDERNESS>> PATRINGTON

184) *ST. OSWALD's* CHURCH, OTTRINGHAM

185) THE ARCTIC CORSAIR [Former Berth on the River Hull - Museum Quarter]

186) STREET LIFE MUSEUM, HULL

187) GANDHI STATUE, NELSON MANDELA GARDENS, HULL

188) BRITANNIA STATUE, NELSON MANDELA GARDENS, HULL

189) THE FROG TRAIL outside Street Life Museum

190) HULL CITY & EAST RIDING MUSEUM, OLD HULL

191) WILBERFORCE HOUSE, OLD HULL

192) BEVERLEY GATE EXCAVATIONS, WHITEFRIARGATE, HULL

193) SIR LEO SCHULTZ BUST, GUILDHALL, HULL

194) [Former] KARDOMAH 94 PERFORMANCE VENUE, ALFRED GELDER STREET, HULL

195) KILNSEA WETLANDS, KILNSEA

196) THE LAST VESTIGE of THE SPURN RAILWAY

197) *CROWN & ANCHOR* PUBLIC HOUSE, KILNSEA

198) *ST. HELEN's* CHURCH, KILNSEA [Closed]

199) KILNSEA's CARAVAN PARK

200) SPURN HEAD VISITOR CENTRE

201) *BLUE BELL INN*, YORKSHIRE WILDLIFE TRUST, KILNSEA

202) EASINGTON TOWER & adjoining METHODIST CHAPEL [Closed]

203) *ALL SAINTS'* CHURCH, EASINGTON

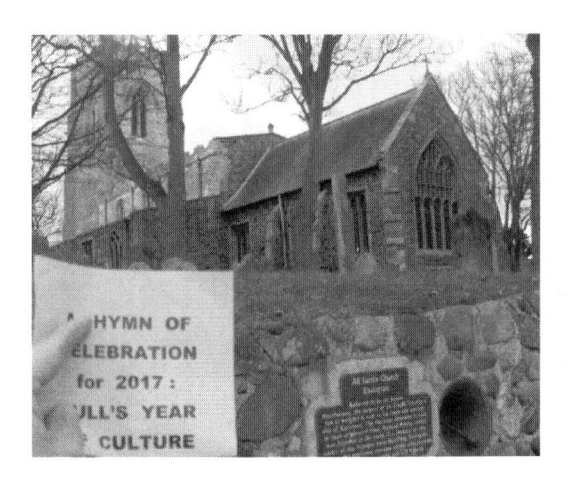

204) WELWICK METHODIST CHURCH / WELWICK COMMUNITY HALL, WELWICK

205) *ST. MARY's* CHURCH , WELWICK

206) HULL KINGSTON ROVERS' RUGBY LEAGUE STADIUM

207) *ST.MATTHEW's* PARISH CHURCH, ANLABY ROAD, HULL [Closed]

208) KCOM STADIUM, HULL FC, ANLABY ROAD, HULL

209) PREMIERE BAR [FORMER CINEMA] ANLABY ROAD, HULL [Closed]

210) CARNEGIE LOCAL HISTORY CENTRE, ANLABY ROAD, HULL

211) *YORK MINSTER*, YORK

212) EAST YORKSHIRE MOTOR SERVICES' BUS DEPÔT, ANLABY ROAD, HULL

213) *THE HUMBER TAVERN*, PAULL

214) PAULL LIGHTHOUSE

215) *TRAFALGAR STREET* CHURCH, BEVERLEY ROAD, HULL [Closed]

216) KINGSTON YOUTH CENTRE, BEVERLEY ROAD, HULL

217) *EMPRESS* PUB, ALFRED GELDER STREET, HULL

218) PARAGON ARCADE, PARAGON STREET, HULL

219) *ST. MARY's* CHURCH, THORNGUMBALD

220) METHODIST CHAPEL, THORNGUMBALD

221) COSTELLO STADIUM, ANLABY PARK, HULL

222) FLYOVER OVER BOOTHFERRY ROAD, HULL

****

A SURPRISING FINISH TO OUTDOOR RECITATIONS OF << *THE HYMN OF CELEBRATION* >>

I had always intended to end my Series of 77 Recitations, on or around December 1st., 2017, on the Humber Bridge where everything started on January 31st.

And, of course, by the end of October, a *new* goal of 222 pop-up Recitations proved achievable.

Even so, on Monday December 4th., I really did travel to the Humber Bridge Country Park. Whilst there, I revisited the Board's Garden of Remembrance, then walked right across the *Eastern* Walkway, recalling 10 months earlier when I had tremulously started out on the *Western* Walkway.

The weather was perfect - as it had been for the great majority of all my excursions whether planned or spontaneous.

Just one omission; one more surprise. On that very last visit to the Portakabin otherwise known as *Humber Bridge Visitor Information Centre*, I came across a Greetings' Card illustrating the Memorial to Hull's Lost Fishermen, St. Andrew's Dock: : a splendid Peter Naylor sculpture I had only ever seen in STAND drawings.

I had no clue, so long after the Hull fishing fleet's triple Disaster, that such a long-delayed, long-negotiated, Project had reached *fruition* .... and how fortuitously, early in 2017: Hull's actual *Year of Culture* ?

And so important was - & is - this outdoor sculpture that I felt compelled to *invent* an extra Recitation [No. 224] in its honour; more surprisingly, a Recitation outside my Book-end January 31st./December 1st.-4th. performance high upon the Humber Bridge.

At least the Memorial is so isolated one of the Humber's forgotten quay-sides, I was not interrupted!

****

CONCLUSION

The challenge for me now is to write a special *Hymn of Celebration* - Commemoration? - the next time there is a special nationally-significant occasion or Anniversary.

A second challenge is for me to take my unique collection of 448 on-location Photographs to Churches & Chapels in both Hull & the East Riding : one timed Photograph showing yellow *Hymn*-sheet at chosen, or accidental, Recitation location; another to show that it was actually *me* there, not someone else.

And there are additional Photographs when I performed a Recitation *indoors* not out. Such opportunities were necessarily much smaller in number, partly because of time constraints; partly because of access; mostly because passers-by might expect the *Hymn* less outdoors than in.

A *third* challenge is for me to pass on everything I have learnt to the next *Year of Culture*: Coventry in 2021; then to the City chosen for PerformingFest *2025*.

What then of those places in the East Riding facing the great City of Kingston-upon-Hull where I did *no* Recitation; where a goal of 333 performances proved unattainable?

I had initially wanted to call in on Goole Docks; the magnificent *Howden Minster*; sturdy, stocky *Holy Trinity Church*, Leven; *All Saints*, Hessle; *Burton Constable Hall* ; *Sewerby*; Bempton Cliffs - the villages of Skirlaugh, Willerby, Flamborough, Brandesburton, wherever.

However, I shall always have the *Hymn* in my pocket when setting off to new places. That's the joy of Hymn-singing, Hymn-learning, Hymn-recollecting: *Amen* is just a four or five verses away.

One morning, the Author sang his Hymn of Celebration within the hallowed walls of St. Patrick's Patrington : the Church for which he wrote his book of Holderness Verse: "Sandcastles do not Fall."

A HYMN OF CELEBRATION for 2017 :

1) GOD BLESS THIS YEAR OF CULTURE :
Of skills, invention, art....
Direct our wills more surely
To play the fullest part.
You gave us joy and anguish :
Hopes dashed, ambitions raised....
Lest we in sloth should languish :
Creation's gifts *be praised*.

2) GOD BLESS THIS YEAR OF MUSIC:
Of solo, chorus, band....
Alert our ears, distinctly,
Each score to understand.
You gave us fine musicians ;
The drum, horn, cello, flute :
Notes chosen with precision;
Performances astute.

3) GOD BLESS THIS YEAR OF PAINTING :
Of landscape, portrait, view....
You gave us oils and pastels ;
Ink, chalk - and subtlest hue.
You nurtured clear perspective :
The illustrative urge ;
Each camera shot elective....
Our visionary surge.

4) GOD BLESS THIS YEAR OF DRAMA :
Of script and costume....*Dance*....
Our love of masquerading;
Deft change of voice and stance.
Each actor's declamation -
Your mission-fields the stage -
Draws loudest acclamation....
Up risen from the page.

                                                    continued........

5) GOD BLESS THIS YEAR OF SCULPTURE
Of object, statue, plate:
Form shaped by flame or chisel,
Symbolic and ornate.
You gave us wood, clay, metal,
The craftsman's steady hold;
Ere final glaze or fettle
Imagination *bold.*

6) GOD BLESS THIS YEAR OF VERSES :
Of poetry and rhyme....
Each tantalizing stanza -
Delivered in strict time;
Quotation loose or metric -
To memorize, recite;
Lines soothing, or electric :
Rehearsed through silent night.

7) GOD BLESS THIS YEAR OF VENTURE :
By dreaming, to conceive ?
Glad days of exultation;
Displays at morn or eve.
You gave us flair and passion
Exciting sense and sight....
Beyond all transient fashion:
Each person's worth, de*light.*

8) *TO HULL, GOD OF ALL CULTURE :*
We varied talents bring:
This year - and in the future -
To serve our glorious King;
So we and all your people -
Blest common purpose found -
May make this chosen City
Your Kingdom's hallowed ground.

****